AF182938

TO OPEN . . .

Confronted with an increasing emphasis on infusing career education into elementary school curricula, teachers frequently find themselves at a loss when looking for career education resources they can integrate into subjects they teach.

Every Number Counts (Level 1) offers rich, diverse opportunities for teaching math concepts and skills through infused career education activities.

MATH AND CAREER EDUCATION

The chief purpose of each unit is to introduce, strengthen, or advance the students' grasp of math concepts and skills. At the same time, each unit focuses on math as applied to the fifteen career clusters authorized by the U. S. Office of Health, Education, and Welfare and adopted widely by schools, school districts, state departments and boards of education, and textbook publishers.

The seven math topics utilized throughout this book represent standard mathematics commonly found in elementary school curricula.

Activities in _Every Number Counts_ have been developed by the authors and have been used with boys and girls in classrooms across America to stimulate and strengthen their awareness that math matters in jobs and careers. And these students have responded with delight and enthusiasm.

FORMAT

Every Number Counts integrates mathematics and career education in a flexible format easy to use. This format includes the following features:

Career Awareness Objectives

Supplemental Objectives (where applicable)

Classroom Activities

Visiting Representative (where useful)

Field Trip (where appropriate)

Dynamic/Interactive Display

Career Awareness Objectives identify the mathematical concept (or concepts) and desired behavioral activity. These objectives also indicate the kind of learning the unit offers: initial awareness, practice, increased understanding, strengthened skills, and other outcomes. Finally, these objectives relate each mathematical topic to appropriate careers and jobs.

Supplemental Objectives identify special career education opportunities related to that topic. These objectives refer to persons from business, industry, and the professions visiting your classroom, sharing their expertise and their enthusiasm for what they do--and usually do so well!

Classroom Activities--in ways the heart of Every Number Counts--describe activities focusing on math concepts and skills and providing experiences and encounters through which students may learn that math does count, indeed, in the work world.

Visiting Representatives, as mentioned above, bring the work world--in rich diversity--to the school world of children, in ways designed to increase their appreciation for math, their grasp of math essentials, their delight in discovering math at work everywhere.

Field Trips describe how to prepare and carry through off-campus experiences which, set in the work world, allow students to meet people using math in their jobs and careers.

Dynamic/Interactive Displays provide plans for three-dimensional displays (thus, dynamic) and bulletin boards created primarily by students (thus, interactive). These object and board displays introduce lively possibilities in environmental learning-scapes, affective growth, and peer learning.

FLEXIBILITY

Authors and artist have designed each unit so that you can dip in and out for ideas, resources, or fully-worked-out activities. If you wish, you can use

EVERY NUMBER COUNTS

LEVEL ONE

BY
RALPH E. KIRKMAN
&
BOB DARRELL
WITH DRAWINGS BY
W.T. WILLIAMS

Library of Congress Card Number 77-82636
ISBN 0-913916-50-1

Copyright © 1977 by Incentive Publications. All Rights Reserved. Printed in Nashville, Tennessee, United States of America. No part of this publication may be reproduced, stored in a retrieval system, or transmitted in any form or by any means, electronic, mechanical, photocopying, recording, or otherwise, without prior written permission of Incentive Publications.

the classroom activity alone; or utilize the visiting representative to follow up on your own lesson; or take the field trip without reference to other features in a particular unit. Too, we believe you will discover that these diverse math and career education strategies will suggest many of your own devising.

EASY USE

To facilitate your use of this book, you will find two appendices composed of (a) detailed listing of career clusters and (b) a chart cross-referencing career clusters, math concepts, and topic summaries. You will find these cross-reference indexes valuable not only for math but also for other subjects.

Every Number Counts contains hundreds of ideas, plans, activities for infusing math with career education concepts. We believe you and your students will enjoy what you uncover and discover.

. . . TeacherTalk

As we mention above, Every Number Counts grows out of the experiments and experiences of hundreds of teachers and students. As you create curriculum activities from our book, we would like to hear how they worked for you. If you make up new activities for math and career education, we would enjoy hearing from you how those projects worked, too. For that purpose, we have included, at the back of our book, a correspondence page entitled "TeacherTalk."

Please write us (our address appears on that page). Let us know what worked and did not work--and why. We will respond. Let us hear from you!

Ralph Kirkman
Bob Darrell

CONTENTS COUNT

SETS
SETS

SETS OF PETS

CAREER AWARENESS OBJECTIVES

Improve ability to perceive sets.

Develop skill at constructing charts.

Increase awareness of how sets help identify animals and products in business--in this instance, the retail pet business.

SUPPLEMENTAL OBJECTIVE

Create opportunity for students to become acquainted with a small business owner and his life- and work-styles.

CLASSROOM ACTIVITIES

Ask students to chart the numbers of various animals, birds, and fish a pet store might sell. Ask them to estimate how much food their pets eat in a week, and how much that food may cost. Ask them to draw a picture of their pet or pets and to write names of their pets below their pictures.

FIELD TRIP

Take your students to a pet shop and let them count number of various animals, birds, and fish (may have to restrict or eliminate fish category because of large quantities). Ask them to work in teams and to count the many different pet foods, leashes, medicines, shelters the store sells.

Ask pet shop personnel to discuss with students how they employ math in their business and to talk about owning a small business.

INTERACTIVE DISPLAY

Display drawings the students did in classroom activity and any posters or items given you by pet store personnel. Students may want to bring photographs of their pets to mount on the display board.

COUNTING NOTES . . .

MISSING MARBLES

CAREER AWARENESS OBJECTIVES

Develop awareness of sets and numbers as part of recreation, sports, and physical exercise.

Learn how math measures exercise, functions in rules for games, and establishes goals in sports and games.

SUPPLEMENTAL OBJECTIVES

Develop awareness of people building careers in recreation and hospitality occupations.

Become aware of lifestyles of recreational and hospitality professionals.

CLASSROOM ACTIVITIES

Draw lines you number and let students play numberline jumping game. Students jump lines and add the numbers to determine their scores. You set a winning score level. (Numerous versions of this game exist; you may want to use your own version.)

Provide a skip rope and let students count the continuous jumps various students complete.

With one or more sets of dominoes, let students create various sets and subsets.

Using six bags, fill each bag with some dominoes (but not all) from the same suit or same colored marbles. Put remaining dominoes or marbles in a sack. Let students team together in two's and try to discover what dominoes or marbles are missing. (If you use marbles, you must tell students in advance that a certain number of marbles should be in the sack. Then, they determine the number in the sack and how many additional numbers are necessary to reach the total you name.) When students identify the missing dominoes or the number of missing marbles, then you let them reach into the remainder bag to retrieve the missing items.

VISITING REPRESENTATIVE

Invite a physical exercise specialist or physical therapist to talk about how he uses math in his occupation, particularly in terms of counting and establishing the number of times an activity is done and rationales for these activities and frequencies.

FIELD TRIP

Visit a local family Y, YMCA, or YWCA, or similar organization to see people working and participating in physical exercise and recreation programs calling for math applications.

DYNAMIC/INTERACTIVE DISPLAY

Place around the room in prominent locations all kinds of exercise equipment and items--particularly ones students can touch.

Let students draw pictures of this equipment; make charts of numbers of lines they jump, numbers of continuous skips of the rope; and other activities. Too, students might draw one of their peers sitting on an exercycle (particularly if you have one on display).

COUNTING NOTES . . .

SEED YOUR SETS

CAREER AWARENESS OBJECTIVES

Develop awareness of role of sets in seed identification and packaging.

Increase perception of growth in measurement terms.

Develop awareness of relationship of math to agri-business occupations.

SUPPLEMENTAL OBJECTIVE

Provide personal meeting with agri-business representatives and their life- and work-styles.

CLASSROOM ACTIVITIES

Distribute various packages of seeds and ask students to count the number of seeds in each package. Then, ask them to create new sets by combining a certain number of seeds from each package. (You can do this same activity with a few packages of seeds by distributing a quantity of seeds--not complete packages--to each student.)

Next, as a continuation of this activity or as a new activity, ask students to plant a certain number of seeds in prepared planting boxes (local gardening stores will help you with this preparation).

This activity will lead to additional set identification and to measurement opportunities as these plants begin to appear. Students then can identify what is growing, can measure at weekly or monthly schedules, and can maintain a chart or graph on growth levels by individual plants and by sets.

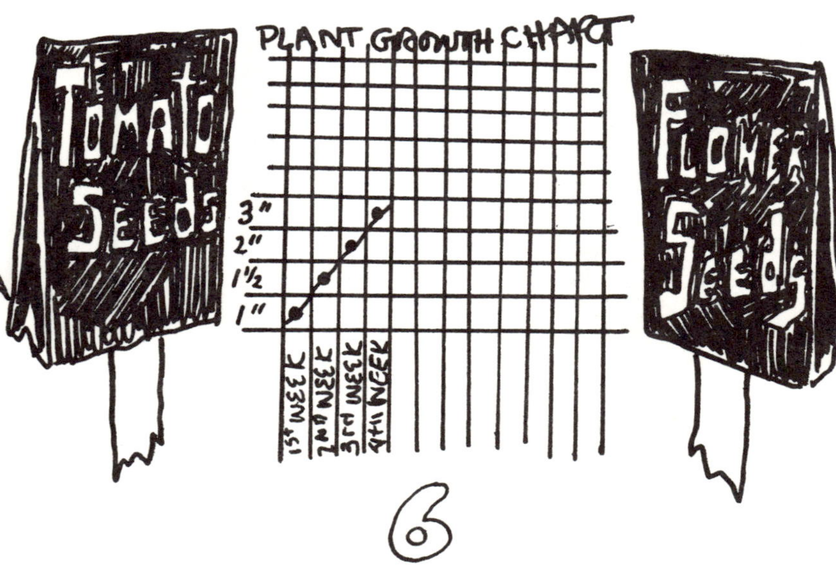

Students can create plant diaries. They draw a picture of the full plant (from the package illustration). Then draw a box under the picture for each week or month. In these boxes they write the date the seeds show growth, place measurement of growth here, and so forth.

VISITING REPRESENTATIVE

Invite a gardener, a retailer in gardening, or a horticulturist to lead the class in activities planned above, if you do not wish to do so yourself. Also, ask him to explain how he utilizes sets and general math in his business.

FIELD TRIP

A visit to a retail gardening store or to a farm allows students to see various items grouped into sets and to practice set reinforcement and identification. Too, this occasion enriches their esthetic sensitivities.

DYNAMIC/INTERACTIVE DISPLAY

With plants growing and growing charts, you have a ready-made, rich display potential. You can saturate the room with these items.

COUNTING NOTES . . .

ONE FOR THE MONEY!

CAREER AWARENESS OBJECTIVES

Develop awareness of role and significance of sets and numbers in marketing and distribution.

Strengthen awareness that people must exchange money--and compute amounts--to buy foods, drinks, and other items sold in such businesses as grocery stores.

Reinforce understanding of sets.

Provide practice in numbers.

SUPPLEMENTAL OBJECTIVE

Provide encounter with grocery store manager to learn how he applies math in his work.

CLASSROOM ACTIVITIES

Direct your students to draw and cut out one $10, two $5, five $1 bills; four halves, four quarters, five dimes, five nickels, and ten pennies. Or, you can make coins out of cardboard, inexpensive checker sets (painted or taped over), or other materials. After completing this activity, introduce sets. For example, discuss the concept that ten pennies equals one dime--that is, these two sets equal each other in money value even though ten pennies creates one set of items and the dime belongs to another set of one item. You may want to bring various empty containers for foodstuffs and let students purchase several items to teach numbers.

Then explain to students how they can join sets; that is, they can put a dime with a nickel and produce a new or mixed set.

Bring a dozen differently labelled cans from home; two or three cans should look alike, and two or three others should carry the same label. With these cans you can let students experiment with creating, joining, and separating sets.

Use pictures of baker, milkman, butcher, produce man, cashier, druggist, and other marketing and distribution persons you wish to add. Multiple copies of these representations allow students practice in sets yet increase their awareness of different marketing and distribution workers.

VISITING REPRESENTATIVE

Invite a grocery store manager to discuss with students the large variety of products available in his store in terms of how he groups them for display, how he classifies them as types of foodstuff and non-foodstuff, and other ways he places them in sets and applies mathematical concepts.

FIELD TRIP

Visit a nearby grocery store. Show students how the store groups products and how many different labels appear in various groups (i.e., six kinds of dog food, eleven types of bread, and so forth). Ask a stocker to discuss his work with students, particularly emphasizing grouping (sets), pricing, and other math matters.

DYNAMIC/INTERACTIVE DISPLAY

Throughout the classroom, build set displays: six apples in a basket in one location; eight cans of peaches in another; four boxes of breakfast food stacked on the floor; and other such displays attractively and strategically placed around the room. (Your grocery store manager will lend you these items; he will be pleased with what you do with math and groceries.) On your bulletin board, attach representations of various marketing and distribution persons--perhaps focusing on employees in grocery and drug stores.

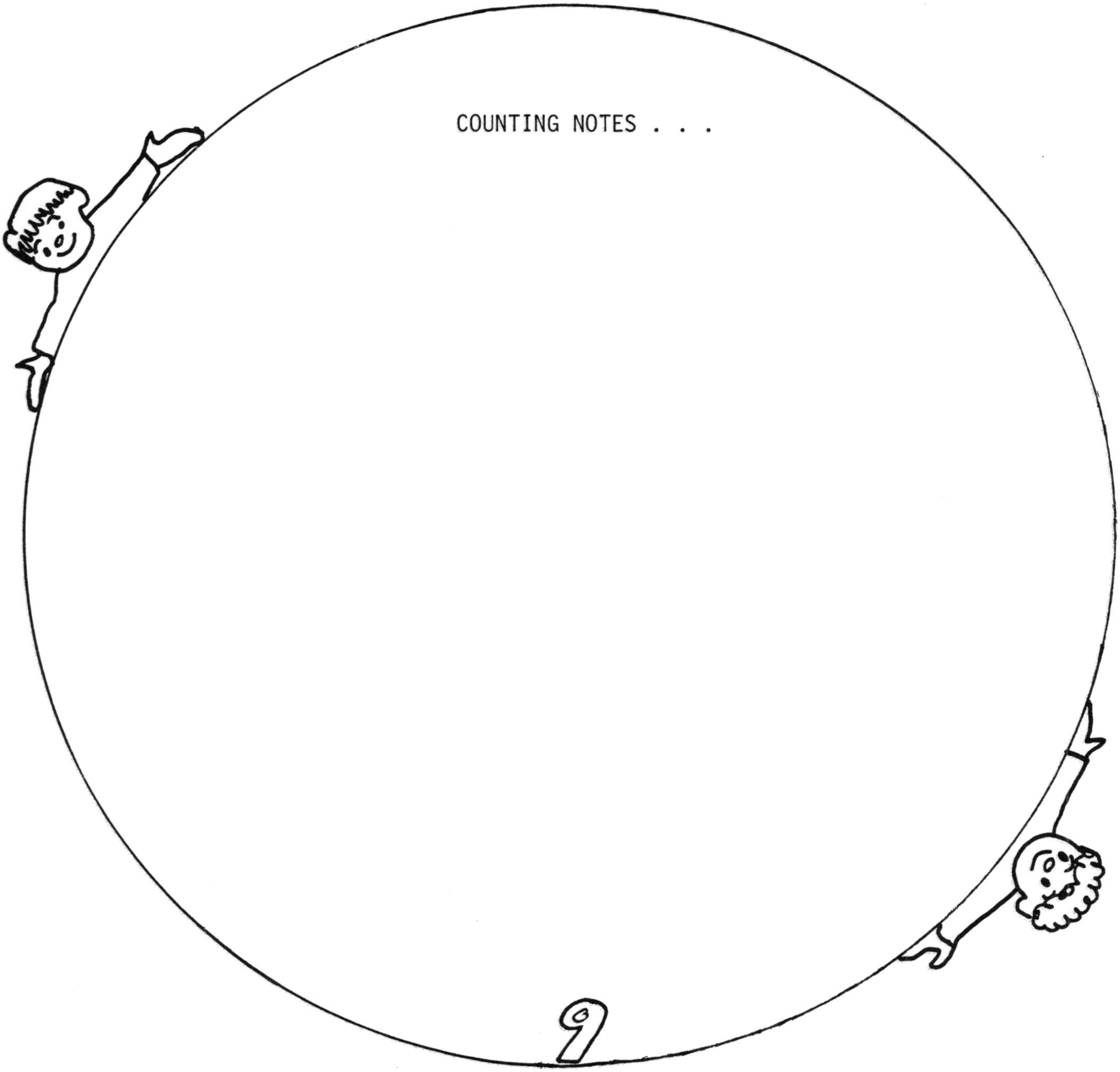

FARM FIGURES

CAREER AWARENESS OBJECTIVES

Increase awareness of sets.

Use sets to expand awareness of agri-business in your community.

SUPPLEMENTAL OBJECTIVES

Provide additional practice with numbers.

Provide association with agri-business representatives.

CLASSROOM ACTIVITIES

Give each student an envelope with a statement inside asking him to locate and cut out as many illustrations as he can of that item. Some items you can use include farm chemicals and fertilizers, farm machinery, farm workers, animals raised for food, animals used for farm work, food crops, clothing crops, and others you may think of. Give students a day or two to locate these items. Then they should bring them back in the same envelope (use 9" x 12" or 10" x 13" envelopes). You, too, should bring several copies of illustrations of each of these items.

Ask students to count how many items they have; let them display them on their desks; you display yours. Next, suggest to students that each one of these desks contains a set. Now suggest that they make up a new set of three items on a subject of their choosing but drawn from three different desks. You can do a number of additional activities with this material, all directed at teaching sets and numbers.

VISITING REPRESENTATIVE

Invite a knowledgeable person to talk about agri-business -- crops & amounts grown, farm support businesses in the area, number of farms, average size, prices, etc.

INTERACTIVE DISPLAY

Allow students to display their cut-out illustrations in a designated area of the room.

SIGNS AND SIGNALS

CAREER AWARENESS OBJECTIVES

Increase awareness of sets.

Enhance awareness of sizes and shapes (geometry).

Increase awareness of variety of traffic signals.

SUPPLEMENTAL OBJECTIVE

Provide opportunity for students to visit with someone working in traffic safety or transportation.

CLASSROOM ACTIVITIES

Ask students to name as many different traffic signs as they can think of. Then, you may want to spark their memories by hinting at ones they do not recall or may not have seen.

Then, ask students to put into one set all signals that suggest STOP; into another set, all signals that indicate YIELD; in a third set, all signals that inform about ROAD CONSTRUCTION; in a fourth set, all signals about turns; in a fifth, all signs about services; and you will think of additional sets or categories.

Next, show students how they might join sets to create specific and actual road situations.

VISITING REPRESENTATIVE

Invite a traffic engineer or an experienced truck driver to talk about varieties of traffic signs, bases on which decisions are made about locating various signs, and so forth.

DYNAMIC/INTERACTIVE DISPLAY

Ask students to draw one set of mixed signs that might be found at a particular intersection--and to include as many signs as would be appropriate. Place these drawings at appropriate places around the room--use some imagination here.

Display various traffic signs lent to you by the community police or traffic control department.

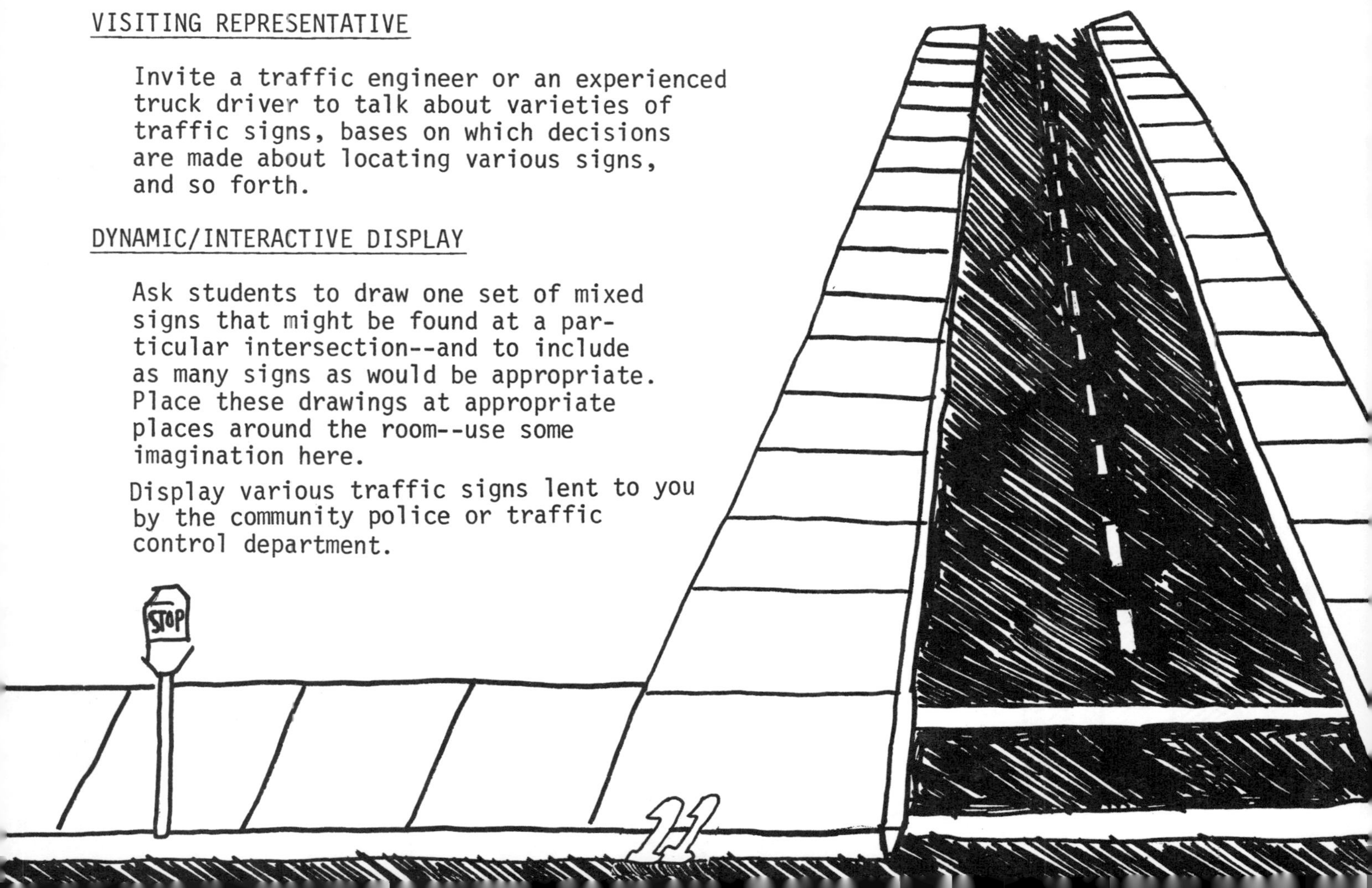

SILVER SETS

CAREER AWARENESS OBJECTIVES

Develop awareness of and distinction between sets, numbers, and numerals.

Learn to construct various kinds of sets of like and unlike items.

Learn to see the relationship between math and homemaking.

CLASSROOM ACTIVITIES

Given silverware for several place settings, the student learns to create place settings, after exercise in counting the numbers of items he has before him. Students establish sets by placing all forks together, all spoons together, etc. Then they count the number of each item.

Next, ask students to create four place settings. They then discover they have created a new set--of place settings rather than of spoons, knives, and forks. Good to have available some serving spoons, steak knives so that students can see possibilities for additional sets. Then, ask students to count objects.

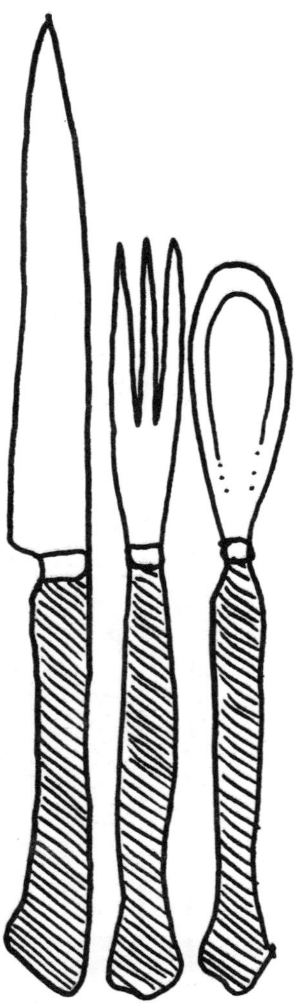

You can do this activity with paper and plastic items and with cups, plates, and napkins--an excellent learning activity for a class picnic as well as a worthwhile classroom activity.

INTERACTIVE DISPLAY

Ask students to draw various place settings and groupings to emphasize sets and numbers of objects. These drawings you attach to your bulletin board. Each day you may want to add some new drawings rather than mount all at one time.

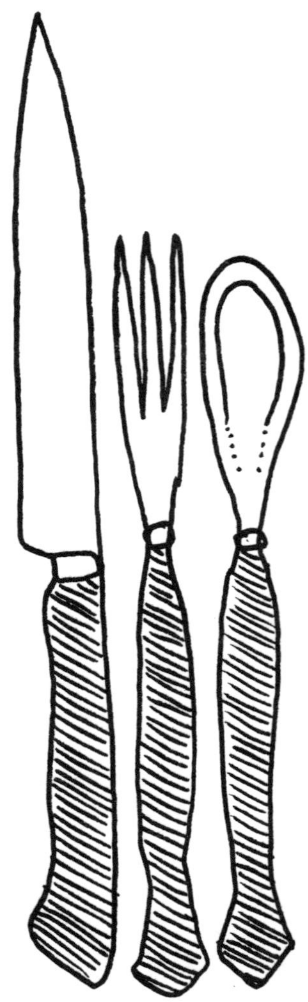

NUMB
COOKIES
CANDY
NEWS

COUNTING CAREERS

CAREER AWARENESS OBJECTIVES

Develop awareness of the function of math in many, many careers besides teachers and accountants.

Increase awareness of great variety of careers to consider.

SUPPLEMENTAL OBJECTIVE

Provide personal meeting with a lively, well-educated person who has had several careers and whom students may have heard about.

CLASSROOM ACTIVITIES

Ask students to list five kinds of occupations or careers they think they would like to work in when they grow up. Then, ask them to indicate which of these probably requires the most knowledge about numbers or arithmetic or math; which, the least. Then ask them, of these two--the one requiring the most math and the one requiring the least--which would they prefer. Then, ask them of the three not included in their last selections, which would they like to do most and which least. Then ask them of these two, which requires more math. You can vary this activity to help students discover careers they find especially appealing--in terms of significance of math skills.

VISITING REPRESENTATIVE

Ask a well-educated person (a good speaker), a man or woman with several careers, to share with the class various math requirements for careers. Let students ask him about various careers they identified in their classroom activity.

INTERACTIVE BULLETIN BOARD

Ask each student to draw a representation of each of the five careers he is interested in and to number on each one first, second, third choice, etc. Place these items on the bulletin board. Also, place any photos you can obtain of the person coming to visit, particularly photos showing him in his various careers.

THE PRICE IS RIGHT!

CAREER AWARENESS OBJECTIVES

Reinforce awareness of numerals as related to but distinguished from numbers.

Intensify skills with whole numbers.

Expand awareness of numerals and numbers in TV, entertainment, and recreation occupations.

SUPPLEMENTAL OBJECTIVE

Provide opportunity for personal encounter with local television and radio personalities to learn how they use numbers in their careers.

CLASSROOM ACTIVITIES

Set up a "The Price Is Right" game for your students. Bring in objects likely to create high interest among your students and ask the principal or some local public personality--if not yourself--to act as master of ceremonies for the occasion. Let students guess at prices, figure how far under or over they are, and provide some inexpensive awards for winners (and a prize for EVERYBODY--even the master of ceremonies and the teacher!).

You might arrange for the master of ceremonies of the TV show, "The Price Is Right," to telephone your class and talk (perhaps via an amplifier--you may borrow one from a local Radio Shack or similar store) to the whole class, letting them ask him a few questions.

VISITING REPRESENTATIVE

The visiting representative would fit into the activity described above. Or, he could describe how numbers, numerals, and whole numbers apply in his occupation--weather, news, ads, payroll, and other ways math would function in TV, radio, and the entertainment business.

INTERACTIVE DISPLAY

Let students draw their conceptions of this show--especially if you have asked them to watch the show before you begin this unit. Then, attach their drawings to the bulletin board.

MEDICAL NUMBERS

CAREER AWARENESS OBJECTIVES

Improve awareness of how numbers figure in keeping and determining time periods.

Reinforce understanding of numbers in scheduling work, prescriptions, and other activities related to allied health and medical professions.

SUPPLEMENTAL OBJECTIVES

Provide practice in whole numbers.

Provide practice in distinguishing numerals and numbers.

Offer opportunity for encounter with allied health and medical professionals.

Enhance awareness of charts and graphs.

CLASSROOM ACTIVITIES

Ask class to set up a hospital, with students serving as doctors, nurses, dietitians, orderlies, and patients. Non-patient personnel establish work schedules, listing times they come to work and leave and establishing shifts so that they cover a twenty-four-hour period. There should always be at least one doctor, one registered nurse, and one orderly available. They all should arrange for a forty-hour work week. They should determine a schedule of how many times they will check patients' well-being, administer temperature readings, deliver pills, and do other routine activities associated with their tasks. The doctor, for example, will make so many in-hospital patient calls daily and will perform a number of operations at certain hours. You may want to arrange to borrow various work and time sheets from your local hospital.

VISITING REPRESENTATIVE

Invite a visiting representative--perhaps a hospital administrator or a supervising nurse to describe how work and treatment schedules are arrived at in a hospital, emphasizing math in his approach. Then, he might evaluate the soundness of the simulation hospital the class has established.

INTERACTIVE BULLETIN BOARD

This bulletin board should display the charts and time schedules students are preparing and be located so that students can modify their displays as they wish.

DOLLAR DISCOUNTS

CAREER AWARENESS OBJECTIVES

Increase awareness of retailing occupations.

Expand awareness of numbers and whole numbers.

Increase awareness of numbers and whole numbers as part of the daily activity of discount or variety stores.

Strengthen skills in using money.

SUPPLEMENTAL OBJECTIVE

Provide opportunity for students to expand knowledge of workers' life- and work-styles in context of retailing industry.

CLASSROOM ACTIVITY

Ask students to make a list of items they would buy at their local discount or variety store. Next, ask them to estimate how much these items will cost. Then, tell them to write down the amount of money in rounded dollar amounts needed to purchase these items. You assign them the amounts they suggest, if these are reasonable. If you do not plan a field trip or visiting representative activity, then come prepared with price list (obtainable from local merchants) for the merchandise they have listed, and check their estimates against the merchants' listing of prices. Those students who have spent more than they had should decide what they will not buy in order to keep their expenses under the assigned amounts.

VISITING REPRESENTATIVE

Invite a visiting representative to come and respond to several of the estimates on various lists. Then, let your visitor answer questions about how he uses math in his work, and let him talk about working in retailing industry.

FIELD TRIP

Take students to discount store or to variety store and let them check prices of items on their lists against shelf prices. Then, if they have spent beyond their limits, they should make adjustments by deciding which items they will remove from their lists of items to purchase.

INTERACTIVE DISPLAY

You can display lists created in the activity or field trip.

CAREER AWARENESS OBJECTIVES

Practice numeral identification.

Recognize functions of numerals in postal work occupations.

SUPPLEMENTAL OBJECTIVES

Develop awareness of at least five different jobs associated with post office activities.

Allow opportunity to visit with postal careers representatives.

CLASSROOM ACTIVITIES

Out of a large carton or box (a furniture store can provide this item for you), construct a post office. Next, locate several uniform small boxes--such as shoe boxes--for children to use as post office boxes. Next, name the aisles between rows of desks as streets: Each student's desk receives the number for his home address on his street. Thus, each desk--including the teacher's--has an address.

Then, ask students to write to each other (perhaps insuring that each person receives at least one letter). All these letters the students mail at the post office, which is then staffed by various working persons. Some of the letters a mailman carries to the addresses; others are kept to be picked up at the post office from the postal boxes. Too, students can sell and buy used stamps (affix with a bit of glue or a tad of transparent tape).

Students practice writing numerals on letters, envelopes; they practice visually recognizing numerals in street addresses on desks. Too, they can count number of letters--practicing numbers. You will think of other activities.

VISITING REPRESENTATIVE

Invite a postal worker, preferably a supervisor, to discuss the ways math functions in his work.

FIELD TRIP

Arrange to take your class to the post office to observe varied operations occurring there, and let them discuss with employees applications of math in postal activities.

DYNAMIC/INTERACTIVE DISPLAY

From your local post office borrow various items and equipment to display throughout the room; borrow posters, too. Also, display letters students write in the activity listed above.

WHAT'S YOUR NUMBER?

CAREER AWARENESS OBJECTIVES

Identify numerals.

Learn to match numerals with appropriate careers.

Reinforce distinction between numerals and numbers.

CLASSROOM ACTIVITIES

From magazines and newspapers, cut out many different illustrations of various careers and occupations: policeman, telephone company, mailman, newspaper boy, milkman, construction (road builder), automobile manufacturer or repairman, service station attendant, airplane pilot or air hostess, weatherman, TV personality, radio personality, and others you may think will work for this activity. Place these illustrations on appropriate, variously-sized cards. Next, draw numbers to suggest these careers: for example, policeman--badge number; mailman--cost of postage stamp; service station attendant--cost of gallon of gas; construction--highway sign number; airplane personnel--AF 1; automobile occupations--license plate; weatherman--temperature; doctor or nurse--98.6^{o}; TV occupations--call numbers of your local TV; same for radio; telephone occupations--telephone number; milkman or newsboy or postman--local residential address; and so forth. These numbers you draw on additional cards. Now you are ready for the activity.

Ask students to match occupation cards with number cards, allowing them to guess broadly within the occupational cluster.

INTERACTIVE DISPLAY

Ask students to find in old magazines similar illustrations and numbers (students can compare these numbers by cutting out various numerals and placing them together to create the numbers they need) and cut them out and bring them for the bulletin board.

Too, you can mount your own placards on the bulletin board.

DENTAL DIGITS

CAREER AWARENESS OBJECTIVES

Reinforce counting skills.

Develop awareness of relationship of math to dentist's activities and dental support careers.

SUPPLEMENTAL OBJECTIVES

Visit with dentist or dentist's assistant.

Develop awareness of life and working styles of dentists and dental assistants.

Create a sense of security about what the dentist does and why he does so.

CLASSROOM ACTIVITIES

Ask students to count the number of teeth in their mouths. Likely, you will receive different totals. That will raise the question "why?" if these students are about the same age.

The dentist can supply you a large model of a set of teeth (perhaps even models of different types) which you can use in counting exercises.

Show students charts of stages various teeth appear (your dentist can supply these charts to you--and other helpful charts). Identify the types of teeth (you can do some work on sets here, if you wish). Let them see how many and what kinds of teeth a person has at one year, at three years, at six and seven years, at ten years, and at fourteen years.

VISITING REPRESENTATIVE

Invite a dentist to talk about how he uses numbers in his work, asking him to illustrate with charts, teeth models, and other supporting materials.

Invite a dental assistant to do the same thing.

FIELD TRIP

Visit a dentist's clinic to see how he works and how he and his dental assistants and secretaries depend upon numbers to accomplish their work. (Perhaps, let students observe the dentist practicing his profession on a patient.)

DYNAMIC/INTERACTIVE DISPLAY

You may want to mount the various charts used in activities described above. Too, you may want to let students draw the number of teeth they have and post these drawings. Your local dental society or a dentist can lend you various items to display.

ADS THAT ADD

CAREER AWARENESS OBJECTIVES

Increase skills in using number concepts and whole numbers.

Expand awareness of widespread role of numbers and whole numbers in various businesses, industries, and professions.

Create awareness of how newspaper industry uses math.

SUPPLEMENTAL OBJECTIVE

Provide opportunity to visit with reporters, advertising persons, editors, classified section people, and other persons associated with the community newspaper.

CLASSROOM ACTIVITIES

Ask students to bring a newspaper with them to class. Then, ask them to go through the paper and cut out every advertisement of a different company that mentions or uses numbers--you may want to ask one student or team of students to concentrate on clothing; another, on automobiles; another, on furniture and appliances; and so forth.

Then ask yet another team to locate news stories which employ numbers in reporting the news.

Discuss the pervasiveness of numbers in these materials. Note that many of these advertisements require ability to add, subtract, multiply, and divide.

VISITING REPRESENTATIVE

Invite a person from the newspaper to discuss how many different ways math figures in the newspaper business.

FIELD TRIP

Take students to the local newspaper office and let them visit with various personnel--always beginning with the question, "How do you use numbers in what you do?"

INTERACTIVE DISPLAY

Cover your bulletin board with newspapers. Then ask students, using magic markers, to draw or write about their experience visiting with the representative or at the newspaper plant.

COUNTING NOTES . . .

NICK NUMBERS

CAREER AWARENESS OBJECTIVES

Increase knowledge of numerals.

Strengthen abilities to recognize distinction between numerals and numbers.

Develop awareness of roles numerals play in identification of public services personnel.

SUPPLEMENTARY OBJECTIVES

Increase recognition of geometric shapes.

Provide personal meeting with public services person.

CLASSROOM ACTIVITIES

Ask students to write numerals for policeman and policewoman's badges, for fireman's hat, for police-car license plate, and for other public-service-connected items you may think of.

Next, ask students to guess how many policemen and policewomen your town employs, how many firemen, etc. Ask students to distinguish between the number of people and the numerals used to identify employees. You may want to send a student to telephone to find out policemen, policewomen, and firemen employee totals.

VISITING REPRESENTATIVE

Invite supervisory-level police officer to confirm student estimates about how many police and fire personnel your city employs. Also, ask officer to discuss ways he uses numerals to identify and to quantify activities and items in his work. Students may want to know, for example, how many prisoners are in the city jail. They may want to know if some kind of numeral system is used to identify these detained persons.

INTERACTIVE DISPLAY

Create a display using numerals to identify persons in public service. Use the same or other numerals to signify numbers of items--try to use pictorial representations of occupational items, activities, and so forth whenever possible. Try to borrow an actual police badge and fireman's badge for your display. You may want to let students who have performed excellently for the day wear the badge for half or whole day the next school day.

BAKER'S BONUS

CAREER AWARENESS OBJECTIVES

Reinforce skills in numbers, numerals, and whole numbers.

Discover math applications in baking industry.

SUPPLEMENTAL OBJECTIVES

Provide personal encounter with a baker.

Gain awareness of baker's life- and work-styles.

CLASSROOM ACTIVITIES

After purchasing all ingredients necessary for baking a cake, let students identify prices, read the recipe and note the math utilized therein, and decide how much change they should receive from a dollar or five dollars for each of the items purchased. Too, let them add up all costs and subtract from a reasonably larger dollar amount--say, ten dollars.

Next, you may want to introduce simple fractions as listed in the recipe.

VISITING REPRESENTATIVE

Invite a baker to come and prepare a cake (or bring one prepared) while he talks about ways he uses numbers in baking and in sales. Too, he may let students do some numerical decorating of the cake prior to letting students eat the cake.

DYNAMIC/INTERACTIVE DISPLAY

Wonderful opportunities here. Ask the baker for display objects and place prices under them. Also, place questions by them asking students how many persons will various items serve, and let students write in answers on the question board.

Display the various ingredients, measuring devices,

and a picture of an oven (maybe an actual oven) around the room. For one day of this activity, you might display doughnuts, cookies, and a pie or cake. Then, invite students to eat them after they engage in a numbers activity involving these pastries.

MAKING MONEY COUNT

CAREER AWARENESS OBJECTIVES

Enhance awareness of math in consumer/homemaking occupational cluster (here, homemaking focusing on personal allowances).

Reinforce understanding of mathematical concepts involved in simple money denominations and transactions.

Sharpen conceptual distinction between numbers and numerals.

CLASSROOM ACTIVITIES

If your students do not have play money they have produced themselves, you may want to purchase several packages of play money, or place play money on sheets and duplicate several sheets of that money which you then can cut out for simulated bills. You will need to create some coins, too--for example, if you save various jar tops, you can paint these to serve as coins. Or, simply cut coins out of cardboard and paint them to look as much as possible like actual coins.

Ask students to think of what they would spend their allowances on. Then, suggest that they pretend they are purchasing some of these items. After they (or you) write down what they decide to buy, they bank with you the money they have spent on these items. Or, they pretend to purchase things they want from each other but actually pay out the money. Further, you can bring various empty containers to class and let students purchase those items as they were marked on the containers.

Ask them to draw various coins and bills (penny, nickel, dime, quarter, half-dollar, dollar) and write in the actual numerals. Ask them to determine how many sets of money they have--can do this with what they draw or with simulated money you make or purchase for them to use for allowances in previous activity.

In discussing allowances, you might want to ask them to consider how much of their allowances they might wish to save and what considerations would lead them to save money.

INTERACTIVE DISPLAY

Students can create huge pieces of money by drawing a bill two feet wide and six feet long (use roll of inexpensive wrapping paper), a penny about a foot in diameter, a dime about ten inches, a nickel about fifteen inches, a quarter about two feet, and a half-dollar about two-and-a-half feet in diameter and use all these items for display. Ask students to bring in any representations of money they find in newspapers or magazines.

CODE ST.

CAREER AWARENESS OBJECTIVES

Reinforce numeral recognition and understanding.

Distinguish between number and numeral.

Increase awareness of the pervasive presence of numbers and numerals in our society--especially in manufacturing.

SUPPLEMENTAL OBJECTIVE

Provide personal visit with a manufacturing representative.

CLASSROOM ACTIVITIES

From various manufacturing firms in your town or county, borrow at least one product. Many products will carry an identification number somewhere on them. Others will show a date of manufacture; some will bear a serial number; still other items will have a numeral or numerals indicating a registration mark (even paper dishes, cups, etc., show such numerals or numbers). Many of these numerals and numbers you will recognize as coding of different kinds; a manufacturing representative can explain most of them to you.

Distribute these products to your students, asking them to locate the numerals and to read them aloud. (You may want to ask them to guess what these numerals mean; give them hints, etc.) Some numerals will refer to quantity of items produced in that series--thus the <u>number</u> of items.

You will think of other activities associated with manufactured items. For example, you might ask each student to bring from home one of his possessions with a numeral on it, and the class can discuss these numerals. Help students see that numerals and numbers are indispensable in manufacturing.

Increasingly, perishable foodstuffs and drug products carry stamped dates indicating dates beyond which a product should not be eaten or otherwise consumed.

You may want to include dates as numbers in your activities.

VISITING REPRESENTATIVE

Invite a senior manufacturing representative to discuss the role of numerals and numbers in business and industry--particularly in relationship to numerals and numbers people see on products.

DYNAMIC/INTERACTIVE DISPLAY

You can saturate your classroom with various products and life-size numerals (three, four, five feet tall!). Place posters sporting numerals and numbers on the board. Also, collect several old license plates and attach them to the interactive board.

COUNTING NOTES . . .

WHOLE
NUMBERS
DEF 3
ABC 2
1
JKL 5
MNO 6
PRS 7
TUV 8
WXY 9
0 OPERATOR

MURAL MADNESS

CAREER AWARENESS OBJECTIVES

Stimulate increased awareness of and interest in types, variety, and number of businesses, industries, and professions in your community.

Provide practice in addition, subtraction, and multiplication.

CLASSROOM ACTIVITIES

Provoke students into brainstorming about the kinds and varieties of careers, jobs, and people who must use numbers and mathematics.

Lead students into sharing what each parent (father and mother) does that uses math--in employment and homemaking.

Direct students to draw a street and locate on that street two, three, or four businesses or companies in their city. After students have drawn these items, take them to the bulletin board and pin these drawings into a large mural. Ask students to sit in front of the mural and to count the number of businesses they see in the mural. Ask them to identify the various kinds of businesses and to add up the total businesses in each category.

Ask them how many of those businesses and industries probably do business with their school.

VISITING REPRESENTATIVE

Ask principal, purchasing agent, bookkeeper, or some other person knowledgeable about who the local school does business with to come in and talk about the school's needs and purchasing in terms of quantities, costs, and budget (could use simplified charts here). Ask person to be prepared to lead students into doing simple addition, subtraction, and multiplication with him as he talks about his subject.

INTERACTIVE DISPLAY

Described under classroom activities.

CARPENTER CAPERS

CAREER AWARENESS OBJECTIVES

Increase comprehension of whole numbers.

Create awareness of role of math--especially whole numbers--in construction activities related to building trades occupations.

SUPPLEMENTAL OBJECTIVES

Provide practice in working with geometrical shapes.

Provide meeting with building trades persons, particularly a carpenter.

CLASSROOM ACTIVITIES

After collecting information on types of materials, dimensions, prices per square foot and per given quantity of nails, and so forth, ask students to compute the cost of building picnic table, dog house, tree house, bird house, set of swings, see-saws, or sand box.

VISITING REPRESENTATIVE

Invite a carpenter to lead this discussion.

DYNAMIC/INTERACTIVE DISPLAY

Display prominently around the room various carpenter's tools and pieces of pipe, wood, and other items related to building trades.

CAREER AWARENESS OBJECTIVES

Increase awareness of math concepts and applications in the communications industry.

Improve understanding of time as a business concept, related to money and sales.

Enhance awareness of and increase practice in addition and subtraction within a business context.

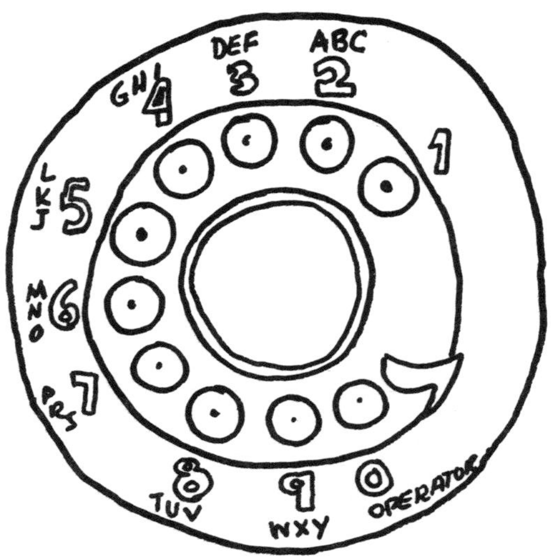

SUPPLEMENTAL OBJECTIVES

Visit with telephone company representative.

Increase awareness of lifestyles of persons working for a telephone company.

CLASSROOM ACTIVITIES

Students can compute a monthly telephone bill based on monthly service charge plus three long-distance calls. Students can compute cost of long-distance call from their hometown to New York City from 8 A.M. to 5 P.M., from 5 P.M. to 11 P.M., and from 11 P.M. to 8 A.M. They then can compare different charges. You may want to suggest other cities to call. (Local telephone representatives will supply you all pertinent data.)

Next, students can call Western Union to find out cost of sending a telegram to New York City and can compare cost with price of telephone call to that city. Too, they can analyze differences in amounts of time needed to complete call or telegram delivery.

Set up an imaginary PBX or telephone switchboard, select a student operator, and let students call through the operator; then, have the operator begin sharing costs of calls to various cities.

VISITING REPRESENTATIVE

Ask telephone company to send representative to visit with your class about how telephone company decides what to charge home customers and business customers, why night calls cost less, and other related matters. This representative also can share with the class how working for a communications company affects his lifestyle.

FIELD VISIT

You can take your class to the telephone company to visit with office personnel, switchboard operators, line crew maintenance persons, and telephone repair service employees. This visit allows students to explore further the relationship between mathematics and the telephone business.

INTERACTIVE DISPLAY

Your local telephone company can furnish posters and brochures for bulletin and wall displays; the company, too, usually will furnish various telephones for display. Students can bring toy telephones from home or old telephones no longer in use. These telephones, plus those furnished by the local company, you can use in activities described above.

Design a large bulletin board to become a simulated monthly billing from the telephone company. Each day, let one student (as a reward for outstanding work that day!) place an imaginary long-distance telephone call on the telephone in the display he likes most. Then he charts the call, the amount of the call, the tax, the date. At the end of the week, students can focus on the billing board to discuss which call costs the most, the least, the total amount of the bill for the class, and other items related to studying whole numbers.

COUNTING NOTES . . .

McNUMBERS

CAREER AWARENESS OBJECTIVES

Increase awareness of math in consumer/homemaking occupational cluster (here, fast-food service)--addition, subtraction, multiplication.

Increase awareness of math in food preparation, packaging, service hours, employment, and buying and selling.

Improve necessary entry skills in math.

Learn to interpret simple recipes.

Reinforce comprehension of sets and numbers.

SUPPLEMENTAL OBJECTIVE

Provide personal encounter with fast-food service visiting representative.

CLASSROOM ACTIVITIES

Using McDonald's Restaurants as the theme, distribute McDonald menus and practice number recognition and set theory: Identify types and number of drinks, sandwiches, pies, cookies. Recognize variations in costs and amounts of differences.

Set up room to simulate a McDonald's Restaurant; let group select waitresses and waiters and customers; let customers order three or four items; waiters and waitresses figure bills (some may make errors); customers check bill to ascertain accuracy. Then some student may order for his family on a take-out basis (multiple orders allow for practice in multiplication on part of customer and waiter).

VISITING REPRESENTATIVE

Invite McDonald's manager or salesperson to class to share about the role of math in his work and lifestyle.

FIELD TRIP

Take students to visit nearby McDonald's (or similar fast-food service outlet) to see how numbers and other aspects of mathematics figure in this business. When students return, discuss what they observed about relationship of numbers to the business.

INTERACTIVE DISPLAY

Children bring various magazine clippings of foods served by fast-food outlets. Place selected ads on bulletin board, already sporting a fast-food service posterboard ad. Then, students guess costs of various items, placing or writing beneath the item two or three "guesstimates." Later ask students to indicate which is higher estimate, which lower, difference between two estimates, and sum of two guesses for each item.

COUNTING NOTES . . .

GREENTHUMBS GROW

CAREER AWARENESS OBJECTIVES

Reinforce perceptions of whole numbers.

Expand awareness of whole numbers in business, especially in a greenhouse operation.

SUPPLEMENTAL OBJECTIVE

Provide opportunity to meet with professional florist, horticulturist, or retail sales person in gardening business.

CLASSROOM ACTIVITIES

Ask principal for a small plot of land in which to grow a small flower/vegetable garden. Six to ten feet should work well. Next, take your students with you to price and buy seeds, fertilizer, and other materials needed (you may want to bring these with you, but let students examine prices, figure out how much you spent, and so forth).

Together plan amounts of watering, watering schedule, fertilizing schedule. Keep a daily log of plants' first appearance. Keep a growth chart (good graphs activity here!).

When plants are ready, you may want to let class organize a plant sale, and sell plants to other students or to themselves, thereby raising some money for next year's garden.

VISITING REPRESENTATIVE

Invite a gardening or florist consultant to share with students about fertilizing and watering and seeding the ground--distances needed to space seeds, and other matters involving math.

DYNAMIC/INTERACTIVE DISPLAY

Students should construct graphs large enough to display on your bulletin board. You may want to place ads about gardening, fertilizer--especially ads and pictures which include matters about math and gardening. Of course, in the growing plants (some of which--particularly if they are herbs--you may be able to transplant for indoor display), you have a marvelous, continuously changing dynamic display!

POP! GO THE MOVIES!

CAREER AWARENESS OBJECTIVES

Provide practice in whole numbers.

Expand awareness of how whole numbers apply in business and careers.

Demonstrate how whole numbers apply in a business associated with fine arts or recreation careers.

SUPPLEMENTAL OBJECTIVES

Help students see the local movies business as requiring knowledge and skill with math.

Provide encounters with persons working in a local movie theater.

CLASSROOM ACTIVITIES

Because of the many occupations involved here--manager, cashier, concessions personnel, ticket taker, ushers, janitors, film technicians, and public relations--you can construct many activities focusing on one or more of these jobs. For example, assume the role of manager and explain that you will have to work out what income you can expect in a week of movies--income derived from films and concessions. Ask students to work with you on some imaginary totals for this week. Then, point out that you have to determine expenditures--food, films, salaries, publicity, and so forth--all of which must be covered by the amount of income expected.

Next, ask students to work out salaries for employees, amount of money to spend for food, amount for film rental, and amount for publicity (newspaper, radio, flyers, etc.).

Now with the basic amounts determined, compare various salaries. Compare amounts spent for food and film to amounts spent for salaries. If expenditures indicate a deficit in light of income, where could students reduce costs, combine jobs, and/or otherwise produce at least a small profit for the theater owner? You will think of many ways to apply whole numbers to this popular entertainment business in your community.

VISITING REPRESENTATIVE

Invite the theater manager to share with students how he uses math--especially whole numbers--in his work.

FIELD TRIP

Take students to the local movie house during a time when many employees will be working but when patrons will not be there (frequently true during weekday school hours). Let various employees discuss how they use math and math concepts in their work.

DYNAMIC DISPLAY

From your local movie house or from your library, borrow (and saturate the room with) movie reels, pieces of film, a theater chair, popcorn sacks and candy wrappers, a movie screen, movie posters and ads, and other items suggesting movies and the movie industry's varied careers.

COUNTING NOTES . . .

PROBLEM SOLVING

DISNEY DELIGHT

CAREER AWARENESS OBJECTIVES

Increase awareness of math in money matters.

Increase skill at solving budget problems with limited monies.

Increase awareness that math is necessary to spend money wisely in recreational and leisure-time activities.

SUPPLEMENTAL OBJECTIVE

Offer students opportunity to visit with amusement park personnel.

CLASSROOM ACTIVITIES

Since Disney World, Animal World, Six Flags Over Texas (Georgia, Mid-America) represent exciting experiences to young persons, you can organize a simulated activity in your classroom. Ask students to set up an imaginary Disney World-- even to coming dressed as Disney characters-- and to establish various rides and amusement events they would like to exist in the classroom. Then, students receive ten dollars, a list of admission prices for each ride, souvenir stand purchase, and foodstuff, and an informal expenses sheet on which they list what they will spend money

on and how much. They then decide how to distribute that ten dollars to provide for one day's entertainment and for adequate food and drink for two meals.

To increase this activity's attractiveness, you may want to ask students who have gone to a Disney World to bring in maps, pictures, stuffed animals, and other objects they have won or purchased at these entertainment centers. (Too, if you will write businesses, they will send you illustrative and billboard materials.)

VISITING REPRESENTATIVE

If your town has an amusement park, invite the owner or manager to visit your class and to discuss how he uses math to operate his enterprise and to solve problems.

FIELD TRIP

An off-season, early, or late season (these times will likely grant complimentary admission and rides--for an educational visit) visit to an amusement park can provide opportunity for your students to practice devising a budget and spending wisely, either actual money or simulated money.

INTERACTIVE DISPLAY

Throughout the room, set up various animals or amusement rides, either in pictorial form or in stuffed-animal or other three-dimensional forms.

WEIGH 'N PAY

CAREER AWARENESS OBJECTIVES

Increase superior students' abilities to solve problems systematically with assistance from mathematical concepts.

Provide practice in dealing with problem solving and logical sequence in the context of a business--transportation.

SUPPLEMENTAL OBJECTIVES

Provide actual contact with shipping and transportation personnel.

Introduce students to actual forms and processes of transportation/shipping procedures.

CLASSROOM ACTIVITIES

Locate a set of scales (at home, nurse's office, or, perhaps, cafeteria) and several boxes filled with any items but weighing one, two, three, four, and five pounds. Ask students to weigh each box and to record weights. Then, suggest they calculate costs for sending each box to a destination at the rate of 50¢ per pound. How much would it cost to send all five boxes?

Since charges for long-distance shipping are set by zones, students can learn to compute costs for shipping parcels long distances by using the following schedule: up to 100 miles--50¢ per pound; 101-200 miles--60¢ per pound; 201-300 miles--70¢ per pound; 301-500 miles--85¢ per pound; 501-750 miles--$1.00 per pound; more than 750 miles--$1.25 per pound.

Next, ask students to transform their figures to metrical equivalents--pounds to grams and miles to meters.

Finally, work with established distances to large cities. With a national road map, determine the miles to selected cities from your hometown and compute costs for variously weighted packages.

FIELD TRIP

Take your students to a shipping company and let one of the company's employees show students the process of receiving and shipping, explaining particularly how problem solving and whole numbers are employed by company workers.

DYNAMIC/INTERACTIVE DISPLAY

Borrow hand-truck, ropes, furniture covers, large invoices, and other items utilized in shipping and display them prominently about the room. Also, you may ask students to draw pictures of various vans and other vehicles transporting goods, and place these on your bulletin board.

COUNTING NOTES . . .

LOTS 'N LOTS

CAREER AWARENESS OBJECTIVES

Learn to solve problems systematically.

Increase awareness of how problem-solving functions as a significant activity in the work world.

Provide practice in solving problems related to a specific occupation--selling a house (real estate).

SUPPLEMENTAL OBJECTIVE

Provide opportunity for students to relate to a real estate broker or sales representative.

CLASSROOM ACTIVITIES

After obtaining from a real estate agency specific information about selling a house and various diagrams and forms the agency would use to gather information about that house, ask students to look at the picture of the house, at internal pictures or diagrams of floor space, closets, and other information. Next, ask students to arrive at guesses about how much floor space, how many closets, how many bathrooms, how many bedrooms, and how many other kinds of rooms and features the house contains. Then, ask them to decide how big the lot should be, how many trees it contains, how big a garage it has, and so forth. Now, they decide what the house is worth. Next, they determine an asking price and a price below which they cannot sell it without losing money. Too, they should decide how much money the agency should receive for the sale and who should pay that fee. After students work up their package, you might let them compare it with the actual package the broker worked out on the house in question. Or, you might telephone the broker (in a conference call) and let students ask him the appropriate questions to determine how accurate, comparatively, the students' package is when compared with what the agency decided.

VISITING REPRESENTATIVE

Invite a broker or real estate sales representative to share his analysis of a given house with that of your class. (This activity, of course, depends on the activity described above.)

INTERACTIVE DISPLAY

Display various ads and pictures of homes taken from magazines, brochures, and newspapers. Also, attach some blueprints and other items the real estate agency supplies.

COUNTING NOTES . . .

DETECTIVE DETAILS

CAREER AWARENESS OBJECTIVES

Improve skills in solving simple arithmetic problems involving schools, a part of the public services occupational cluster.

Increase awareness of how regularly and frequently teachers, maintenance personnel, librarians, and principals must apply simple arithmetic knowledge and skills to solve problems.

Reinforce skills in numbers and numerals.

Reinforce knowledge of measurement concepts.

Reinforce understanding of whole numbers.

CLASSROOM ACTIVITIES

Give students a ditto of their school building with rooms drawn in according to location. Ask students to write in the numbers of rooms as they remember them. When they have finished, let students ask one another for room numbers they do not remember, thus allowing students to fill in as many room numbers as they collectively can remember. You may want to send out "detectives" to scout for the missing numbers.

Ask students to count the number of desks in their room, number of books, of chairs, and other multiple items.

Ask them to determine how many additional chairs they would need if one parent for each child visited their room.

Ask students to count how much money they spend for lunches in a week.

Ask students to count the number of letters in their names.

Ask students to determine how many persons can sit in their school bus.

Ask students to count the number of days they actually go to school, the number of days until George Washington's birthday, and the number of days until school is out in the spring, and other such calendar activities.

Ask students to count the number of teachers in the school.

INTERACTIVE DISPLAY

Create a bulletin board with various visual representations of items to be counted. Allow space under, beside, or above for answers to the questions implied in your group presentations. Also place question marks at appropriate places on the board. Students fill in these spaces as they discover and confirm the answers.

COUNTING NOTES . . .

DIVING FOR NUMBERS

CAREER AWARENESS OBJECTIVES

Expand awareness of role of math in solving problems.

Create awareness of math in marine science occupations.

Reinforce perception of and skills in numbers.

SUPPLEMENTAL OBJECTIVES

Reinforce perception abilities in sets.

Reinforce skills in working with whole numbers.

Provide opportunity to meet with persons working in marine sciences or with persons highly knowledgeable about these occupations.

CLASSROOM ACTIVITIES

Since marine scientists study the characteristics, movements, and properties of bodies of water and the immediate environments of those bodies, you can do the following activity: Take students to a nearby river, lake, pond, or ocean. Ask them to examine closely the kinds of living things they find--variety, number, size, shape, and other characteristics. Ask them to consider what kinds of bad things could happen to disrupt the body of water and the environment as they have just found it. Ask them what kinds of things could they and other people do to preserve this environment. Keep in mind that problem solving involves identification of the problem, determining alternative solutions, and decision making which results in solution.

VISITING REPRESENTATIVE

Invite a marine scientist or a local expert in natural waters in your area to describe local natural water conditions and perhaps pollutions. Ask him to describe how professionals study marine life and activity and how they attempt to solve problems.

FIELD TRIP

The classroom activity described above is a field trip, but you can enrich it by inviting a marine scientist to lead it.

DYNAMIC/INTERACTIVE DISPLAY

Collect various kinds of underwater equipment--including diving, measuring, and transportation. Place this equipment around the room (local sports stores, students, and parents will lend you these items). From magazines cut out various illustrations of marine life activity and occupations. Place on bulletin board. Ask students to draw pictures in two- or three- or four-panel sequences in which they introduce a problem and solve it.

COUNTING NOTES . . .

SPORTING PROBLEM

CAREER AWARENESS OBJECTIVES

Increase ability to solve problems with assistance of math.

Develop awareness of relationship among money, small business, and math concepts.

SUPPLEMENTAL OBJECTIVE

Provide opportunity for students to become acquainted with small business owner, specifically a sports store owner.

CLASSROOM ACTIVITIES

First, on large mural paper (or a mattress or bicycle carton), paint a sporting goods store (or store front). Next, provide students with play money (they can make their own). Then ask them to bring sports items from home, for which they make price tags (have some actual tags for them to use; see a local store for these). Now, students go to the shop to purchase items for sale. They get to play with what they buy for a given time. To begin the actual buying process, give each person an envelope with a problem inside. For example, one such problem may be to buy three items with $25, or with $10. They have to decide what they can buy, how much money remains, and other activities.

A variation on this activity involves rules books for several sports. The mural would carry drawings of various referees, umpires, and so forth for the major sports--and different shapes of balls and other items used in the games. This variation does not need the play money but does need several rules books for sports. The envelopes carry problems which students can answer by consulting the rules books. Remember that you want each problem to involve math in the solution.

VISITING REPRESENTATIVE

Invite a sporting goods store owner or manager to share with your students the kinds of problems he solves with math.

DYNAMIC/INTERACTIVE DISPLAY

Ask students to bring at least one cut-out of a sporting event for the bulletin board. Display prominently throughout the room the various items students bring. You may want to talk to the physical education teacher or the coach for lending you some equipment, too!

MEASUREMENT
1 2 3 4 5 6 7
3 7/8"

METER MAGIC

CAREER AWARENESS OBJECTIVES

Increase awareness of the role of mathematics in utilities occupations.

Improve necessary entry skills in gas, electric and water meter reading.

Increase skill in employing math to measure.

Enhance awareness of math, particularly how math measures utilities necessary to home and business.

CLASSROOM ACTIVITIES

Show simulated drawings of water, gas, and electric meters, and describe how to read them. Then, ask students to divide into groups of three or four and show them how to figure usage (visiting representatives can come to do this for you, if you wish). Next, ask student groups to establish usage for a month of water, of electricity, of gas. Next, give them the basic rate and ask them to multiply to determine actual cost of consumption for each during the month. (You can plan this activity so that students can bring data from their home bills to increase realism and relevance of mastering this skill and of knowing this information.)

VISITING REPRESENTATIVE

Ask a city employee--either a meter reader who communicates effectively or a public relations person or supervisor--to visit and discuss what he does with these readings, who takes the figures and computes bills, how rates are arrived at, when peak usage periods occur, and related items. He may want to illustrate readings from an actual model meter (which you also can display).

FIELD TRIP

Visit a utility company to allow students to see employees at work, to see computer facilities, to ask questions about math's role in the business, to learn how these people feel about working for the public and as public service persons. You may want to prepare students by giving them sample questions to spark their questions related to measurement activities in utilities. Also, you can lead your class to various meters located on your schoolground and let them read these instruments.

INTERACTIVE DISPLAY

In addition to various measurement devices you display, ask local utilities companies to supply dummy meters and other measurement devices used in their activities.

Prepare a chart or graph allowing students to maintain a month's readings on electricity and water and gas consumed at their school. At month's end, let them compute costs for the period. (They could do the same for home, and check to see how close they come to actual bill.)

COUNTING NOTES . . .

WEIGHT WORKS!

CAREER AWARENESS OBJECTIVES

Increase skills at recognizing varying weights.

Increase awareness of relationship of career occupation to workplace.

CLASSROOM ACTIVITIES

Obtain a simple balance and different measures of weights (available in many school offices, post offices, and businesses), and assure yourself that these weights clearly and accurately indicate the actual weights. Collect pictures of different places of employment and of the people who work at them. Next, fasten a weight on the back of the picture of the person and an equal weight on the back of the corresponding workplace. Thus, the person and the place will balance, be appropriate, and signify each other.

INTERACTIVE DISPLAY

Ask students to draw several scales balancing career person with workplace. Also ask them to draw some career persons and workplaces which do not correspond, giving them opportunity to draw scales out of balance.

MIND YOUR LINES!

CAREER AWARENESS OBJECTIVES

Expand awareness of linear measurement.

Realize importance of reading measurements accurately.

Develop awareness of relationship of measurements to construction.

SUPPLEMENTAL OBJECTIVE

Provide encounter with personnel in construction, architecture, and related fields.

CLASSROOM ACTIVITIES

Divide students into two, three, or four groups. Give each group materials to construct a simple box form--giving some students correct measurements, and some incorrect measurements. Check results. Most likely, some you have given correct measurements to will misread or miscalculate. Others will follow carefully both the correct and incorrect measurements. For materials, you can use paper, wood, and so forth. This unit can apply to working with fabrics, too.

VISITING REPRESENTATIVE

Invite a builder, construction worker, architect, or some other person involved in using measurements to construct an object. Ask him to talk about the function of accurate measurement and of measurement generally in his occupation.

3 7/8"

4 1/8"

FEET COUNT! METERS MATTER

CAREER AWARENESS OBJECTIVES

Increase awareness of measurement.

Expand perception of various athletic playing areas as constructed with aid of math and measurement.

Reinforce notion of importance of accurate measurement.

Practice conversion to metrics.

SUPPLEMENTAL OBJECTIVE

Provide meeting with a coach or prominent local sports personality or athlete and increase awareness of math in his work or leisure-time activity.

CLASSROOM ACTIVITIES

If you have a football, baseball, or basketball area, take students out to measure it (borrow from real estate office, engineering firm, or local government several long metal tape rules and other types of equipment used for measuring considerable distances). Ask students to measure the field(s) and to convert these measurements into metric measures as well.

Ask students to consider the various amounts of playing time involved in major sports. Too, ask about how playing time is divided. Ask students to compare the various amounts of time and how the time is divided. Are total game times about the same? Are methods of dividing time about the same?

VISITING REPRESENTATIVE

Invite a coach, athlete, or prominent sports personality to discuss the role of math in sports. This discussion need not be narrowly related to measurement to be highly effective.

DYNAMIC/INTERACTIVE DISPLAY

Ask students to bring various types of sports equipment and various measuring devices applicable to measuring in sports. Then group these in displays about the room, including at least one measuring device in each display. Ask students to draw their favorite game and to try to draw the field as accurately as they can (you may want to suggest scale here if you believe your students ready for simple scale work: for example, 1" for 50').

WAGES'N GAUGES

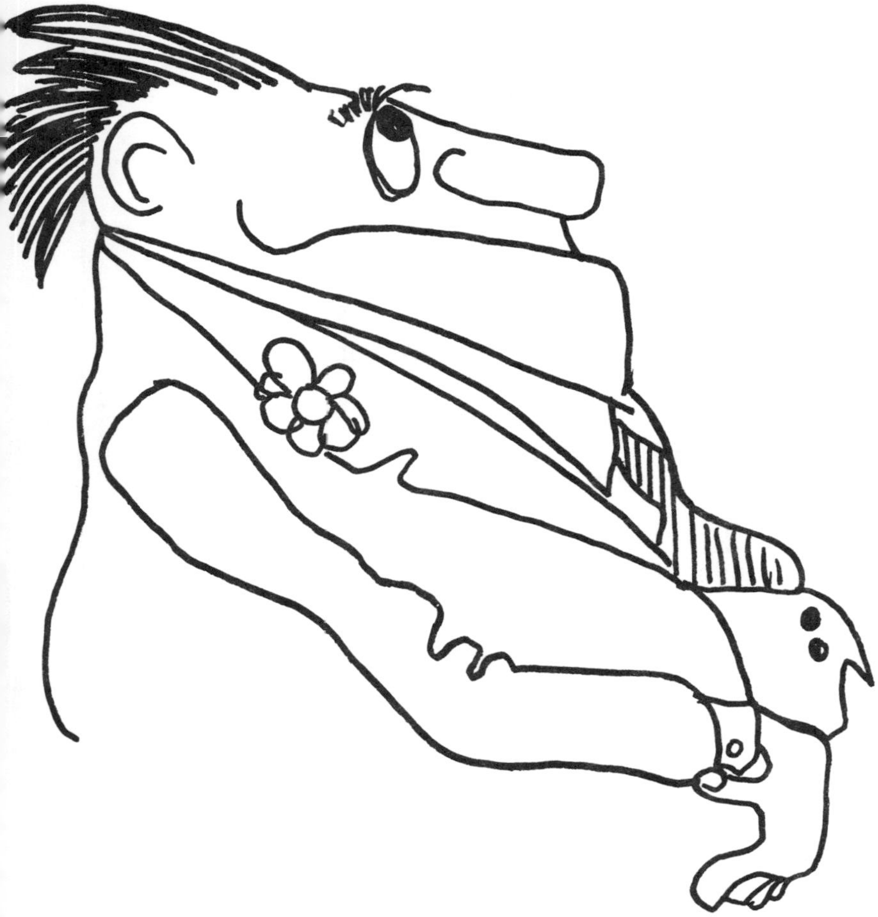

CAREER AWARENESS OBJECTIVES

Enhance awareness of math in a wide range of career clusters--measurement, sets, numbers.

Stimulate increased awareness of and interest in types, variety, and number of businesses, industries, and professions in community.

Provide practice in measurement, number identification, and sets.

Increase skills in identification of various measurement devices.

CLASSROOM ACTIVITIES

Ask your class to brainstorm to identify workers using math in their daily jobs--bankers, cashiers, bookkeepers, pharmacists, engineers, pilots, chemists, astronauts, teachers, weathermen, and scientists.

Direct students to bring a wide variety of measuring devices for display purposes. Each student should prepare a list of possible career applications for each device. The class should discuss these measurement devices in terms of applications to careers.

VISITING REPRESENTATIVES

Ask several children's parents who represent businesses, industries, or professions utilizing measurement devices to visit your class to discuss math uses of these devices in their careers and lifestyles. At least three to five careers should be represented in the discussion.

INTERACTIVE DISPLAY

Display on a saturation basis, placing equipment and items (e.g., yardsticks, thermometer, barometer, timer, simple cash register, clock, abacus, gauges, weights, containers) throughout the room--shelves, floor, etc. Encourage students to touch and examine items (hands-on activity they can do on a browsing basis and as reward when other work completed).

PERTINENT PERIMETERS

CAREER AWARENESS OBJECTIVES

Increase awareness of measurement in construction and in town zoning.

Increase awareness of measure in occupations devoted to building trades and to city planning.

Increase awareness of lifestyles of construction people and of public officials.

Increase awareness of use of charts and graphs in city planning.

Provide practice in metrics.

Provide practice in applying measuring concepts.

SUPPLEMENTAL OBJECTIVE

Create opportunity to meet with personnel in building trades and city government.

CLASSROOM ACTIVITIES

Ask students to measure distance between two locations within the school.

Next, ask students, working in pairs, to obtain one of the following measurements: (1) the classroom's perimeter, (2) classroom area, (3) amount of carpeting needed to carpet the room, (4) amount of space, approximately, covered by desks, (5) blackboard area, (6) door areas, (7) window areas, (8) book and cabinet area, (9) distance of room exits to building exit (closest doors), (10) and other pertinent measurements.

Then ask students to convert these yards and feet measurements to metric measurements.

Now students can draw a scale model of the classroom using the information gathered by the teams (some should work in yards and feet; others in meters and centimeters). Scale: one inch for one yard; one centimeter for one meter.

Distribute city or area zoning maps. Ask students to make graphs illustrating amounts of city devoted to commercial, to residential, to public (parks, schools, etc.), and to farm areas.

VISITING REPRESENTATIVES

Invite a construction worker to discuss his trade, particularly referring to how he employs measurements in his various tasks.

Invite a city engineer or planning commission chairperson to discuss the role of mathematics in city planning, zoning, and other planning activities.

FIELD TRIP

Visit a construction site--particularly one with several units under construction--to allow students opportunity to measure on-site construction.

INTERACTIVE DISPLAY

Post on bulletin board scale drawings completed by students. Also, place various measuring (linear) devices around the room and on the board. Ask students to watch local paper for news items and for advertising which uses measuring information. They should bring these clippings to attach to your bulletin board.

COUNTING NOTES . . .

TREES TREES TREES

CAREER AWARENESS OBJECTIVES

Suggest applications of math, especially measurements, to environmental occupations.

Provide practice in measurements, numerals, numbers, and whole numbers.

SUPPLEMENTAL OBJECTIVE

Visit with nursery manager or horticulturist.

CLASSROOM ACTIVITIES

Create on a bulletin board or on a large carton (such as a mattress or bicycle carton) a tree with fully-grown branches and large roots. Let students suggest occupations which help plants grow, and write in on each root the name of an occupation which helps plants or foods grow. On each limb, write in names of occupations that benefit from natural objects such as trees. Next, count up these occupations, etc.

Then, take students outside to observe and measure trees of varying sizes--heights, circumferences, various sizes of leaves, and so forth.

You may want to bring a book that discusses various sizes different trees reach at maturity.

FIELD TRIP

Take students to a nursery center which sells trees. Ask the manager to help students learn about different trees, measure the small seedlings available for sale, and learn about the differences in current heights and circumferences and mature heights and circumferences.

INTERACTIVE DISPLAY

Ask students to locate various pictures of trees in magazines and newspapers, and let them attach these pictures onto bulletin board or carton displaying the big tree mentioned in activity listed above.

GEO

SUPERSIZES, MARKETSHAPES

CAREER AWARENESS OBJECTIVES

Increase awareness of sizes and shapes.

Deepen recognition that items given the same money value may vary in sizes, shapes, and weights (colors, too).

Increase awareness of roles of personnel working in a supermarket or large grocery store--various specialists.

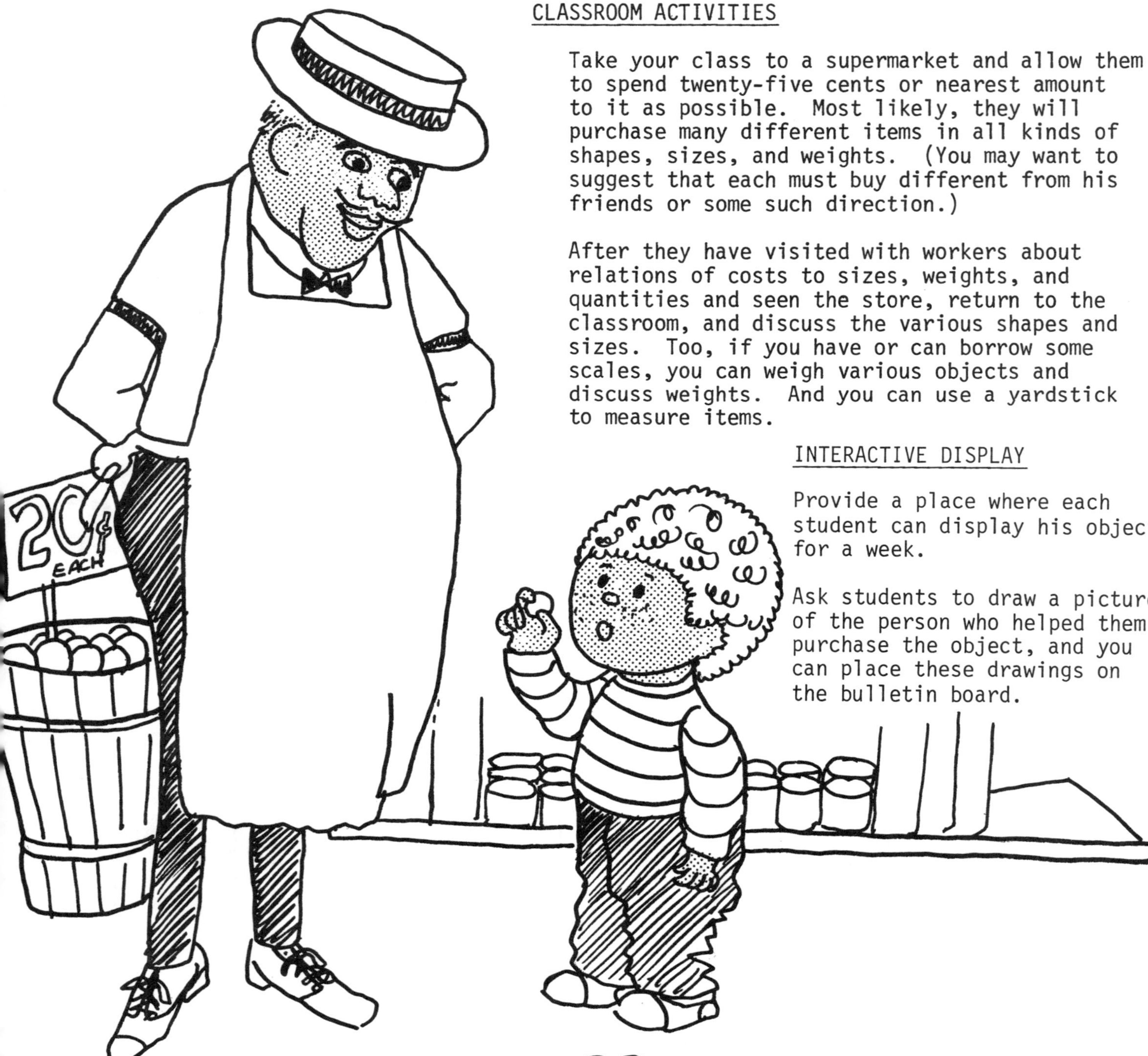

CLASSROOM ACTIVITIES

Take your class to a supermarket and allow them to spend twenty-five cents or nearest amount to it as possible. Most likely, they will purchase many different items in all kinds of shapes, sizes, and weights. (You may want to suggest that each must buy different from his friends or some such direction.)

After they have visited with workers about relations of costs to sizes, weights, and quantities and seen the store, return to the classroom, and discuss the various shapes and sizes. Too, if you have or can borrow some scales, you can weigh various objects and discuss weights. And you can use a yardstick to measure items.

INTERACTIVE DISPLAY

Provide a place where each student can display his object for a week.

Ask students to draw a picture of the person who helped them purchase the object, and you can place these drawings on the bulletin board.

MATCHING MODELS

CAREER AWARENESS OBJECTIVES

Expand awareness of shapes and sizes.

Explore relationship between shapes and sizes and various modes of transportation.

Consider relationship of shapes to ease or speed of movement.

CLASSROOM ACTIVITIES

From photographs you make, magazine and newspaper cut-outs you obtain, or from posters, display shapes of various modes of transportation. Ask students to consider how these shapes suggest conventional geometrical shapes. You may want to make illustrations of these shapes and encourage students to match geometrical shapes with shapes of various types of transportation or even parts of transportation vehicles.

DYNAMIC/INTERACTIVE DISPLAY

Ask students to bring to class different toys suggesting geometrical shapes and representing the transportation field. Display prominently these contributed items. Local bus companies or branches, airline terminals, train stations, and some auto dealers will lend attractive models, too.

You may want to distribute conventional geometric designs, some of which do not appear in transportation vehicles, and ask students to draw transportation vehicles of the future and for the future. Then, post these drawings near your displays of current transportation vehicles.

SHAPE UP!

CAREER AWARENESS OBJECTIVES

Develop awareness of the presence of basic geometric forms in uniforms, in items used for work, and in human figures in public service occupations.

Create or intensify recognition of geometric forms.

Increase perception of various shapes and forms.

SUPPLEMENTAL OBJECTIVE

Provide personal encounter with public services professionals--policeman or policewoman, fireman, sanitation collection personnel, and city engineer.

CLASSROOM ACTIVITIES

Ask students to draw a figure of a police officer using geometric shapes, calling attention to the uniform--including badges on coat and hat, traffic ticket book, and billy club. Also, ask students to create a figure of a sanitation collection person holding a garbage can and lid, again creating through geometric shapes.

Describe duties of a traffic or street engineer. Then ask students to draw a street light, a traffic light signal, various traffic signs, a street, walk, and curb--all based on appropriate geometric shapes.

VISITING REPRESENTATIVE

Invite the city engineer to describe ways he employs geometric forms in his work.

INTERACTIVE DISPLAY

Begin a bulletin board with a few cut-outs of items from newspapers which obviously utilize simple geometric forms. Then ask students to bring such geometric shapes they find in newspapers and other publications. Too, they should try to bring occupationally-oriented objects embodying various geometric shapes--certain tools, clothing, sports equipment, and other materials and objects. Moreover, you can arrange furniture, equipment, and books in a room to emphasize geometric forms.

SET THE STAGE!

CAREER AWARENESS OBJECTIVES

Increase recognition skills regarding shapes.

Improve whole number skills.

Create awareness of how whole numbers and sets figure in planning a dramatic production.

SUPPLEMENTAL OBJECTIVES

Create opportunity for students to visit with director, actors and actresses, and production staff.

Provide some reinforcement of measurement skills.

CLASSROOM ACTIVITIES

Ask how many theatres and movie houses in your area. Classify them into sets. Estimate how many people can attend each one and all together. (You may want to collect this information in advance for yourself.) Compare how many can attend drive-ins, movie houses, and play houses.

Discuss the shapes of these buildings outside and inside, the lay-out of seating, etc. Discuss the various kinds of personnel needed to maintain the various kinds of theatres. (Most of your students likely will have attended one or more of these types of theatres.)

VISITING REPRESENTATIVES

Invite a director, actors and actresses, and some production personnel--all these may be working with a high school, college, or civic production--to discuss how they utilize math in preparing a production.

BONUS: Perhaps, the visiting contingent will perform a scene for your class!

FIELD TRIP

Take your students to the most interesting theatre in your area. Let them count seats, lights, curtains (nearly always more than two!), and other countable items. Let them figure cost of admission and add and subtract and multiply for various money denominations and for group admissions. Let them figure the area of the stage and the seating area. You will think of other applications when you see the building from the perspective of math applications.

DYNAMIC DISPLAY

Borrow commercial poster displays, props, autographed photos, and other items related to movies and theatres. Post these around the room in prominent, unusual locations.

COUNTING NOTES . . .

CHARTS,
GRAPHS, &
STATISTICS

CHART YOUR HEART!

CAREER AWARENESS OBJECTIVES

Develop awareness of mathematical concepts in charting a patient's progress, relapse, and subsequent progress.

Intensify awareness of how medical and allied health professionals utilize math concepts in their work.

SUPPLEMENTAL OBJECTIVE

Increase awareness of lifestyles of medical and allied health professionals.

CLASSROOM ACTIVITIES

Following classroom visitation or visiting representative activities described below, create a large patient progress chart which allows students to keep temperature records, pulse rates, and other statistical medical information they wish to chart. For example, students can record pulse rates for selected students for a week and chart the rate pattern (work with school nurse here if you wish), or they can maintain records on pulse, temperature, etc.

Also, students should chart when their patient receives pills (and the number of different pills), when he should receive shots, how many X-rays he should undergo and when he should do so, and other items associated with his general treatment that are susceptible to charting, graphing, and statistical representation.

VISITING REPRESENTATIVE

Invite nurse or doctor to describe how each keeps charts or graphs of patient activity. These visitors also can describe other ways they use graphs and charts in their professional activities.

FIELD TRIP

A field trip to a clinic or hospital to meet with several staff representatives and to view charts and graphs would benefit students, allowing them to see directly the relevance and significance of such math skills and enabling them to see the work and lifestyles of medical and allied health professionals. Properly prepared, this activity can introduce students to a variety of charts and graphs (some perhaps projected on screens).

WEIGH IN

CAREER AWARENESS OBJECTIVES

Increase awareness of how math measures physical growth.

Develop awareness of numbers as ways to measure growth in inches, metrics, and pounds.

Reinforce perception of distinction between number as a quantity and numeral as the character signifying that quantity.

CLASSROOM ACTIVITIES

Upon students' entry into class, or shortly thereafter, measure and weigh students. Then chart results. You can do this on a huge linear chart running along one wall and dropping to the floor. At far right, mark September (or the current month). Then every two or three or four months, again measure and weigh students, recording the results and writing the month at the far left--at the level generally reached by the class.

You can help students learn metrics, too, if you chart these statistics in inches, feet, ounces, pounds, AND centimeters, meters, decagrams, and kilograms. You may want to add metrics later and go back and convert the earlier measurements.

You can dress up this activity with attractive seals, insignias, objects with adhesive backs to stick at the points children reach in linear growth.

INTERACTIVE DISPLAY

Of course, the above acts as a display. You can substitute, however, a smaller chart or graph, which you can place on your bulletin board. You may want to let students write in their own statistics.

CHECK IT OUT

CAREER AWARENESS OBJECTIVES

Develop awareness of how charts help people understand concepts, routines, and problems.

Think of charts as an activity related to occupations, particularly to a librarian's career.

SUPPLEMENTAL OBJECTIVE

Provide opportunity for students to visit with a librarian as he shares how he uses math in his work.

CLASSROOM ACTIVITIES

Help students develop questions that you consider chartable regarding a librarian's work. Such questions can include the following: How many hours daily do you work? How many books are checked out each day? How many books do you keep in science, how many in fiction, how many in math, how many in career education, how many in reference, how many in music and art? How many books do you add each year? During a five-year span? How much money do you spend on book purchases, on magazines, on bookbinding, on other library activities? You will think of many additional questions students can ask that will provide information they can chart. Next, show students how to create simple charts. You may want them to guess at answers or you may want to provide them answers, which they can chart.

VISITING REPRESENTATIVE

Invite your school librarian to share his answers with students. Too, he may want to tell them how much time he spends working, caring for family, pursuing recreational activities, and eating and sleeping. The librarian also can provide a work break-down indicating how much time he spends checking out books, how much time shelving, how much time keeping records, how much time attending meetings, and how much time apportioned to other work functions.

If your students interview the librarian, you can divide them into teams to work up charts. The result of their work they might share with the librarian.

INTERACTIVE DISPLAY

Students can attach to the bulletin board the charts they create. Also, you might cut out various charts from magazines and newspapers to enhance this display.

CHURCHES 'N CHARTS

CAREER AWARENESS OBJECTIVES

Increase ability to chart facts and statistical data.

Expand awareness of how charting enables people to make judgments about work loads and other activities connected with jobs and careers.

Provide practice in charting and graphing.

SUPPLEMENTAL OBJECTIVE

Provide meeting with several ministers.

CLASSROOM ACTIVITIES

These classroom activities depend upon visiting representatives. First, help students develop questions they wish to ask visiting clergy, questions which will provide chartable data. Such questions may include the following questions: What is the size of your congregation (in families or in individual persons)? How many persons attend Sunday church school--how many youth, how many adults? How many persons do you visit in the hospital in a given week? How many home visits do you make in an average week? How many meetings do you attend in a typical week? How many sermons do you preach at your church in a year? How many baptisms do you administer annually? How many weddings do you perform? You will think--and so will your students--of many, many other questions which will provide good chartable data and will reveal the nature of the minister's work.

VISITING REPRESENTATIVES

Invite at least three--better four?--ministers, priests, rabbis, and other full-time religious workers to share this interview with students. Remember you need to have primed your students with questions and you may have to play secretary (or use a tape recorder) to record the pertinent data for the graphing and charting activities which will follow the interview.

After the clergy visit, show students how to graph and chart the data. Help them to interpret the meaning of the data they do chart.

INTERACTIVE DISPLAY

Place charts on bulletin board. You may want to invite another class to see the board since many of them will belong to some of the churches and synagogues represented in the interview.

CHECK'N STUB

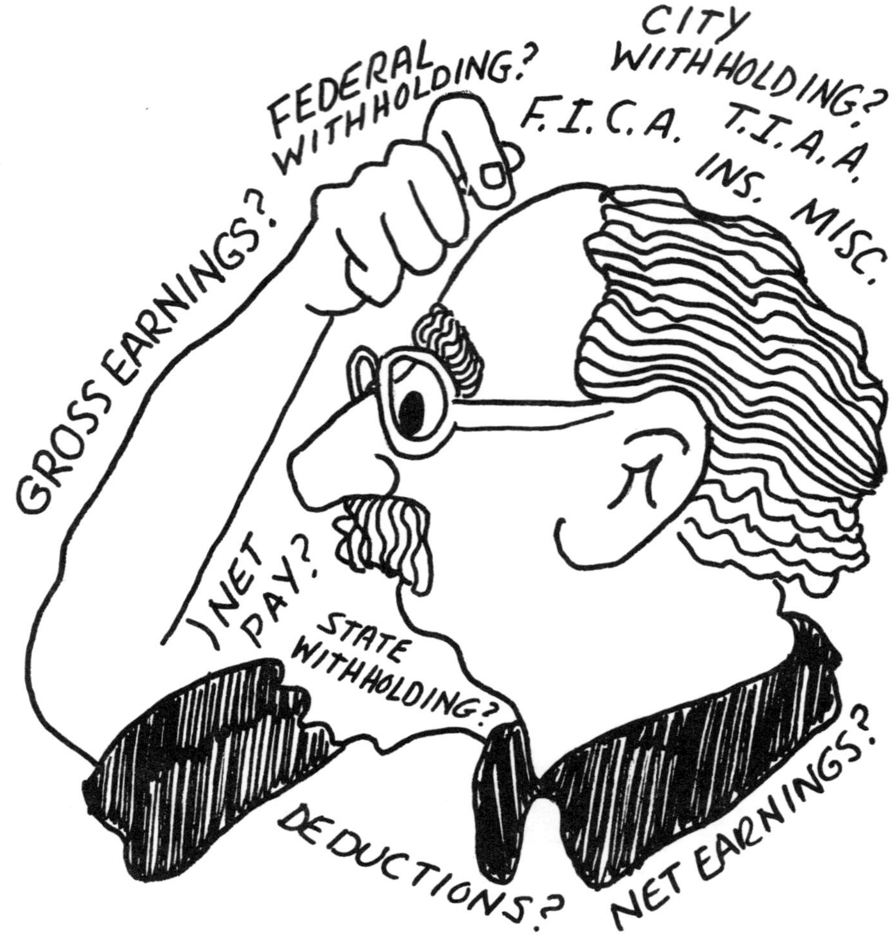

CAREER AWARENESS OBJECTIVES

Increase understanding of how charts work.

Expand awareness of charts in banking, business, and office occupations.

SUPPLEMENTAL OBJECTIVES

Provide practice with numerals, numbers, and whole numbers.

Provide encounter with business representatives and their working and lifestyles.

CLASSROOM ACTIVITIES

Create a large chart of a typical payroll check illustrating gross salary, payroll deductions, and net salary (most businesses will provide you examples--some will offer you samples to distribute to your students). With this large chart, discuss differences between gross and net income, kinds of payroll deductions and methods of deductions, destination of monies deducted, and other related items.

Create the same chart with a check and its stub, showing how to fill in a check properly, how to enter and subtract amounts, how to check for accuracy.

VISITING REPRESENTATIVE

Ask business or banking representative to come in and perform activities described above. Or, ask that person to share with your students ways he uses charts in his work.

FIELD TRIP

Go to a business or a bank and let students see the workplace, ask questions of working personnel, receive demonstrations of various ways math--especially charting--are used in that business.

INTERACTIVE DISPLAY

Ask students to draw their own payroll check, fill in figures, and determine gross salary, various deductions, and net salary. Then, place these drawings on the bulletin board. Also, you can obtain from the business community or a bank various financial statements and items to display.

COUNTING NOTES . . .

BARGAIN HUNTER

CAREER AWARENESS OBJECTIVES

Introduce bar and line graphs.

Strengthen perception of role of math in work world, particularly in retail sales industry.

SUPPLEMENTAL OBJECTIVES

Provide meeting with a comparison shopper to see how this worker uses math and graphs in his work.

Increase awareness of the variation in merchandise costs and the importance of comparison shopping as a consumer.

CLASSROOM ACTIVITIES

Frequently, comparison shoppers report their findings on graphs using bars and lines. Set up a problem such as the following: a large supermarket manager wants to check the costs of a quart of milk, a pound of apples, and a pound of hamburger meat in stores competing with him. So, he sends the comparison shopper to gather that information (and you supply the students with five prices for each of these products). Then, you say to students that we need some way of seeing these comparisons easily, clearly. Then, construct a model graph. Next, ask students to construct a model graph for the quart of milk prices, then for the pound of hamburger meat, and so forth.

Next, ask students to figure out how much more or less the competitors were selling each product for compared to the supermarket manager's prices. Then, ask them to figure which store sold all three items at the lowest price, which at the highest, and where the supermarket manager's prices fall in comparison.

VISITING REPRESENTATIVE

Invite a comparison shopper or a supermarket manager to share how comparison shoppers work and why, with particular emphasis on math concepts such as bar and line graphs.

INTERACTIVE DISPLAY

Design a huge bar and line graph for your room--cover a complete wall. On your bulletin board, place the graphs your students make. Using pictures or drawings of foods utilized in the activity above, write in the various prices you use in the comparison shopping activity.

COUNTING NOTES . . .

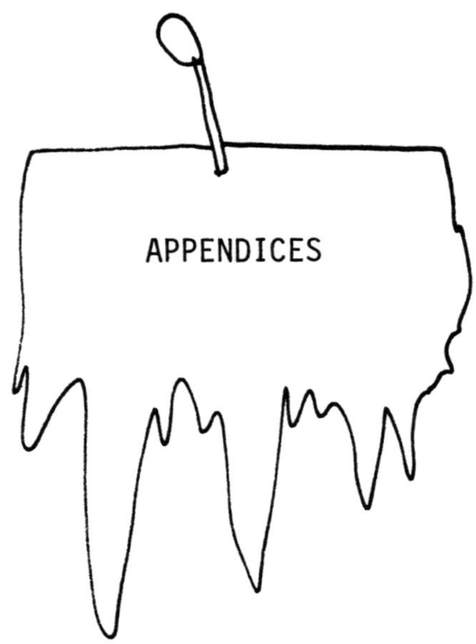
APPENDICES

APPENDIX A

CAREER CLUSTERS

I. AGRI-BUSINESS AND NATURAL RESOURCES

Animal Husbandman
Animal Research Scientist
Cooperative Extension Service Worker
Cowboy
Entomologist
Farm Worker
Farmer--dairy, livestock
Forester
Geologist
Geophysicist
Microbiologist
Miner
Oil Refinery Worker
Petroleum Engineer
Plant Research Scientist
Rancher
Rig Builder
Soil Conservationist
Soil Scientist
Veterinarian
Veterinary Aid
Wild Life Scientist

II. BUSINESS AND OFFICE

Accountant
Bookkeeper
Clerk Typist
Computer Programmer
Dealership Operator
Fiscal Manager
General Office Clerk
Insurance Agent
Keypunch Operator
Machine Transcriber
Office Manager
Office Supervisor
Real Estate Broker
Secretary
Stenographer
Systems Analyst

III. COMMUNICATION AND MEDIA

Creative Writer
Darkroom Technician
Darkroom Worker
Display Technician
Editor
Electronics Engineer
Illustrator
Journalist
Layout Man
Photographer
Reporter
Sign Painter
Telephone Installer
Telephone Lineman and Cable Splicer
TV and Radio Announcer
TV and Radio Engineer
TV and Radio Program Director
TV and Radio Repair Technician

IV. CONSTRUCTION

- Architect--commercial
- Architect--landscape
- Architect--residential
- Bricklayer
- Carpenter
- Cement Mason
- Construction Laborer
- Electrician
- Elevator Constructors
- Engineer--civil
- Engineer--electrical
- Engineer--mechanical
- Floor Covering Installer
- Glazier
- Iron Worker
- Lather
- Metal Worker
- Painter
- Paper Hanger
- Plasterer
- Plumber and Pipefitter
- Roofer
- Sheet Metal Worker
- Stonemason

V. CONSUMER AND HOMEMAKING

- Chef
- Cleaner
- Clothing Assembler
- Clothing Cutter
- Clothing Designer
- Clothing Inspector
- Clothing Machine Operator
- Clothing Pattern Maker
- Clothing Presser
- Clothing Weaver
- Consumer Counselor
- Cook
- Dietician
- Food Journalist
- Food Processor
- Food Purchaser
- Food Service Industry Worker
- Food Tester
- Furniture Maker
- Furniture Refinisher
- Home Economist
- Home Management Counselor
- Household Day Worker
- Housekeeper
- Interior Decorator
- Laundry Worker
- Presser
- Product Demonstrator
- Upholsterer
- Waiter

VI. ENVIRONMENT

- Air Analyst
- Anti-pollution Law Enforcer
- Bacteriologist
- Conservationist
- Engine Emission Inspector
- Exterminator
- Farmer
- Florist
- Game Warden
- Gardener
- Hydroponic Plant Culturist
- Industrial Pollution Controller
- Landscape Architect
- Marine Biologist
- Meteorologist
- Mine and Dump Restoration Controller
- Pollution Control Engineer
- Pollution Regulation Enforcer
- Sanitation Men
- Soil Analyst
- Urban Planner
- Water Analyst

VII. FINE ARTS AND HUMANITIES

Actor
Artist Manager
Cameraman
Costume Designer
Costume Maker
Creative Writer
Dancer
Fashion Designer
Film Editor
Film Processor
Film Producer
Foreign Language Broadcaster
Illustrator
Industrial Designer
Minister
Music Composer
Painter
Printmaker
Sculptor
Stage Lighting Designer
Stage Manager
Stage Set Constructor
Tape Editor
Tape Producer

VIII. HEALTH SERVICES

Biochemist
Biophysicist
Chiropractor
Dental Assistant
Dental Hygienist
Dentist
Dietician
Electrocardiographic Technician
Hospital Administrator
Inhalation Therapist
Licensed Practical Nurse
Medical Assistant
Medical Laboratory Worker
Medical Photographer
Medical Records Librarian
Optometric Assistant
Optometrist
Osteopathic Physician
Pharmacist
Physical Therapist
Physician
Podiatrist
Radiologic Technologist
Registered Nurse
Sanitarian
Veterinarian

IX. HOSPITALITY, RECREATION, AND LEISURE

Bellmen and Bell Captain
Bus Driver
Chef
Church Group
Coach
Front Office Clerk
Hotel or Motel Manager
Movie Projectionist
Museum Curator
Museum Guard
Park Maintenance Worker
Park Planner
Park Ranger
Playground Supervisor
Professional Athlete
Senior Citizen Center Supervisor
Sports Director
Theatre Usher
Tour Guides
Travel Agent
Waiter
Waitress

X. MANUFACTURING

Assembly Line Operator
Distributor
Equipment Producer--Foundry
Equipment Producer--Machine
Industrial Engineer
Industrial Psychologist
Plant Manager
Product Designer
Product Developer
Product Inspector
Product Packager
Product Promoter
Product Promotion Researcher
Product Storer
Product Tester
Product Transporter
Retailer
Wholesaler

XI. MARINE SCIENCE

Deep Sea Diver
Fish Hatcher and Raiser
Marine Animal Researcher
Marine Biological Laboratory Worker
Marine Plant Researcher
Ocean Fishing Boat Operator
Oceanographic Mapper
Ocean Mineral Explorer
Offshore Mineral Driller
Offshore Pipe System Constructor
Seafood Inspector
Sport Fishing Processor
Underwater Construction Worker
Underwater Demolition Worker
Underwater Engineering Researcher
Underwater Salvage Operator

XII. MARKETING AND DISTRIBUTION

Clothing Store Owner/Manager
Department Store Owner/Manager
Drug Store Owner/Manager
Gas Station Attendant
Grocery Store Owner/Manager
Management Personnel
Marketing Researcher
Pet Shop Owner/Manager
Purchasing Agent
Salesman
Sales Promotion/Publicity Personnel
Variety Store Owner/Manager

XIII. PERSONAL SERVICES

Animal Boarder
Animal Groomer
Animal Trainer
Barber
Bartender
Building Custodian
Chef
Dry Cleaner
Guard
Hair Stylist
Laundry Worker
Makeup Technician
Masseur
Private Household Worker
Waiter
Watchman

XIV. PUBLIC SERVICE

Agricultural Advisors
Army
Building Inspector
Coast Guard
Counselor--Rehabilitation
Customs Officials
Elected Officials
Employment Counselors
Engineers
Firemen
Hospital Workers
Immigration Inspectors
Internal Revenue Service
Judges
Librarians
Maintenance Workers
Marine Corps
Navy
Plant and Animal Inspectors
Policemen--City or State
Port Authority Workers
Post Office Workers
Public Recorders
Registrars and Licensers of Cars
Research Workers
Sanitation Men
Teachers
Traffic Controllers

XV. TRANSPORTATION

Aircraft Mechanics
Airline Dispatchers
Air Traffic Controllers
Automobile Mechanics
Bus Drivers
Clerks
Conductors
Flight Engineers
Locomotive Engineers
Locomotive Firemen
Merchant Mariner Officers
Pilots
Railroad Brakemen
Signal Workers
Station Agents
Stewards and Stewardesses
Taxi Drivers
Telegraphers, Telephoners, Tower Men
Track Workers
Traffic Agents
Truck Drivers
Truck Mechanics
Unlicensed Merchant Seamen

APPENDIX B

CAREER CLUSTERS, MATH CONCEPTS, AND TOPIC SUMMARIES

CAREER CLUSTER	MATH CONCEPT	TOPIC	PAGE
Agri-Business and Natural Resources	Sets	Seed Your Sets	6
	Sets	Farm Figures.	10
	Whole Numbers	Greenthumb Grow	40
Business and Office	Problem Solving	Lots 'N Lots	48
	Charts, Graphs, and Statistics	Check 'N Stub	76
Communication and Media	Numbers and Numerals	The Price Is Right	16
	Numbers and Numerals	Ads That Add	25
	Whole Numbers	A Time to Call	36
	Charts, Graphs, and Statistics	Check It Out	74
Construction	Whole Numbers	Carpenter Capers	35
	Measurement	Mind Your Lines!	59
	Measurement	Pertinent Perimeters	62
Consumer and Home-making	Sets	One for the Money!	8
	Sets	Silver Sets	12
	Numbers and Numerals	Dollar Discounts	20

Books in the KIDS & CAREERS Series:

EVERY NUMBER COUNTS - Level 1 by Darrell & Kirkman

EVERY NUMBER COUNTS - Level 2 by Darrell & Kirkman

AT LEAST A THOUSAND THINGS TO DO by Farnette, Forte & Loss

I'VE GOT ME AND I'M GLAD by Farnette, Forte & Loss

TEACHERTALK

Unit name____

What worked well?

What did not work well?

Recommendations regarding future use of this unit:

Please share with us other math and career education activities you considered successful:

Your Name_____________________________

Address_______________________________

City, State, Zip______________________

____

SEND TO:

Ralph Kirkman
Bob Darrell
Every Number Counts
Incentive Publications
2400 Crestmoor Drive
Nashville, TN 37215